EROSION

PROPERTY OF ORLAND SCHOOL DISTRICT

Ellen Ripley

New York

Published in 2009 by The Rosen Publishing Group, Inc.
29 East 21st Street, New York, NY 10010

Copyright © 2009 by The Rosen Publishing Group, Inc.

All rights reserved. No part of this book may be reproduced in any form without permission in writing from the publisher, except by a reviewer.

Book Design: Daniel Hosek

Photo Credits: Cover © Mary Lane/Shutterstock; p. 5 © Wolfgang Lienbacher/Stringer/Getty Images; p. 6 © Anton Foltin/Shutterstock; p. 7 © Michael Hare/Shutterstock; pp. 8–9 © Julián Rovagnati/Shutterstock; pp. 10–11 © Rémi Cauzid/Shutterstock; p. 11 (top) © Hulton Archive/Stringer/Getty Images; pp. 12–13 © Sally Scott/Shutterstock; p. 14 (background) © Liset Alvarez/Shutterstock.

ISBN: 978-1-4358-0101-1
6-pack ISBN: 978-1-4358-0102-8

Manufactured in the United States of America

CPSIA Compliance Information: Batch #WR313190RC:
For Further Information contact Rosen Publishing, New York, New York at 1-800-237-9932

Contents

What Is Erosion? — 4

Water Erosion — 6

Ice Erosion — 8

Wind Erosion — 10

Erosion Caused by People — 12

Glossary — 15

Index — 16

What Is Erosion?

Erosion is the breaking up and movement of Earth's land. Landslides are one example of erosion. A landslide happens when part of a **slope** breaks away and **gravity** pulls the loose earth down the slope.

Several forces can cause erosion. Over many years, wind and water can cause erosion. Ice and gravity can also cause erosion. Sometimes the things people do cause erosion where it doesn't commonly happen.

This landslide happened in April 2008 in Salzburg, Austria. It blocked a small road and almost blocked the river, too.

Water Erosion

Did you know that water can erode land? This **process** often takes many years. Ocean waves crash onto shores again and again. Over time, the waves break the ground into rocks. The rocks are slowly broken down into tiny bits of sand.

Flowing rivers dig into the earth and make **riverbeds** wider and deeper. This action fills the rivers with tiny pieces of earth called **silt**. The silt builds up at the ends of rivers, often making new land.

The Colorado River, shown here, eroded the earth to make the Grand Canyon.

After many years, the water and silt in this waterfall have eroded the rocks below.

Ice Erosion

Water sometimes fills tiny breaks in rocks. When it gets cold, the water freezes and forms ice. The ice causes the breaks to grow larger. After this happens many times, the ice can snap rocks apart.

A **glacier** is a large body of slow-moving ice. It grows larger as more snow falls on it. After many years, a glacier may begin to slide down a mountain slope or across land. This erodes the land beneath the glacier.

This is a glacier in the Andes Mountains of Argentina.

Wind Erosion

Wind can also erode land. Wind picks up very tiny pieces of rock and sand and blows them against larger rocks. Over time, this process wears away rocks and mountains. It makes them smoother or gives them interesting shapes.

Wind erosion can be a big problem for farmers in hot, dry places. When there is nothing to block the wind, it can blow away much of the soil on a farm. Then crops have trouble growing.

This picture, taken in 1894, shows a dust storm rolling over a farm in Midland, Texas.

Wind erosion helped create these land formations in Monument Valley, Arizona.

Erosion Caused by People

Trees help keep erosion from happening in forests. Tree roots hold soil in place and help keep it tight and firm. The branches and leaves slow the fall of rain and keep it from eroding the soil.

Clearcutting, shown here, is a form of logging that removes all trees from a part of forest.

Logging can cause problems in forests. Removing trees can lead to erosion. Even the logging roads add to the problem. They speed up erosion. People have to be careful not to cause erosion where it doesn't commonly happen.

Main Topic
Poor logging practices can cause erosion in forests.

Fact
Poorly planned logging roads can cause landslides.

Fact
Without tree roots, forest soil more easily breaks apart and washes away.

Fact
Without tree branches, more raindrops hit the soil harder. This can increase erosion over time.

Fact
Rain can't go into packed soil on logging roads. Instead, it runs off, taking soil with it.

Erosion

What It Is

The breaking up and movement of Earth's land

Causes

- gravity
- water
- ice
- wind
- people

Examples

- waves crashing on shore
- glaciers sliding over land
- wind blowing tiny pieces of rock and sand
- landslides

Problems Caused By Erosion

Logging can cause erosion in forests where it wasn't a problem before.

Glossary

glacier (GLAY-shur) A large buildup of ice that moves slowly over land.

gravity (GRA-vuh-tee) The natural force that causes objects to move toward Earth.

logging (LOG-ing) The act or business of cutting down trees.

process (PRAH-sehs) A set of actions that happen in a certain order.

riverbed (RIH-vuhr-behd) Ground covered by a river.

silt (SILT) Fine bits of earth carried away by rivers.

slope (SLOHP) Land that rises upward.

Index

F
forces, 4
forest(s), 12, 13, 14

G
glacier(s), 8, 14
gravity, 4, 14

I
ice, 4, 8, 14

L
landslide(s), 4, 13, 14
logging, 13, 14

M
mountain(s), 8, 10

P
people, 4, 13, 14
process, 6, 10

R
rain(drops), 12, 13
riverbeds, 6
rivers, 6
rock(s), 6, 8, 10, 14

S
sand, 6, 10, 14
silt, 6
snow, 8
soil, 10, 12, 13

W
water, 4, 6, 8, 14
waves, 6, 14
wind, 4, 10, 14

Due to the changing nature of Internet links, The Rosen Publishing Group, Inc., has developed an online list of Web sites related to the subject of this book. This site is updated regularly. Please use this link to access the list: http://www.rcbmlinks.com/rlr/erosi